The Scientific Method
IN ACTION

GUESS!
RESEARCH AND FORM A HYPOTHESIS

Emma Carlson Berne

PowerKiDS press.
New York

Published in 2015 by The Rosen Publishing Group, Inc.
29 East 21st Street, New York, NY 10010

First Edition

Editor: Jennifer Way
Book Design: Kate Vlachos
Photo Research: Katie Stryker

Photo Credits: Cover, p. 13 Blend Images/Thinkstock; p. 4 Monkey Business Images/Shutterstock.com; p. 5 Sergey Novikov/Shutterstock.com; pp. 7, 10 (right), 14, 20 iStockphoto/Thinkstock; p. 8 Stephen Simpson/The Image Bank/Getty Images; p. 9 Photos.com/Thinkstock; p. 10 (left) kali9/E+/Getty Images; p. 11 Tom Wang/Shutterstock.com; p. 15 Pressmaster/Shutterstock.com; p. 17 ZouZou/Shutterstock.com; p. 18 Fuse/Thinkstock; p. 19 vovol/Shutterstock.com; p. 22 Popperfoto/Contributor/Getty Images.

Library of Congress Cataloging-in-Publication Data

Berne, Emma Carlson.
Guess! : research and form a hypothesis / by Emma Carlson Berne. — First edition.
pages cm. — (The scientific method in action)
Includes index.
ISBN 978-1-4777-2926-7 (library) — ISBN 978-1-4777-3014-0 (pbk.) —
ISBN 978-1-4777-3085-0 (6-pack)
1. Science—Methodology—Juvenile literature. I. Title.
Q175.2.B472 2015
507.2'4—dc23
2013025379

Manufactured in the United States of America

CPSIA Compliance Information: Batch #WS14PK5: For Further Information contact Rosen Publishing, New York, New York at 1-800-237-9932

CONTENTS

SEARCHING FOR KNOWLEDGE

These scientists are looking at samples under a microscope in order to test their hypothesis.

Scientists are always searching for knowledge. You do not have to wear a white coat and work in a **laboratory** to be a scientist, though. When you ask questions and look for answers, you are a scientist, too!

Scientists search for answers to their questions by doing **experiments**. They use a plan called the **scientific method**. The scientific method has six parts. These parts are making **observations**, forming a **hypothesis**, planning an experiment, collecting information, **analyzing** the results, and sharing the results.

We will learn how to form a hypothesis in this book. To do this, we will focus on **researching topics** and making educated guesses.

You do not have to wear a white coat to be a scientist, but you do have to be safe. Wear safety goggles and rubber gloves when experimenting.

OBSERVING AROUND YOU

The first step in creating a hypothesis is thinking of a question you want to answer. The question should come from observations that you have done. You can choose to observe something that you have noticed in your daily life. You may have noticed that when you have a glass of cold water on a hot day, the outside of the glass becomes foggy and wet. The same thing does not happen on a cool day, though.

Next, form the question you want to answer. Your question should be **specific** and one you can explore by doing research. Your question could be: Why does the outside of my glass of ice water become wet on hot days?

These children are thinking of scientific questions while observing the world around them. They might wonder what makes the pond water look murky.

WHAT DO YOU KNOW?

If you are interested in a topic that involves things you cannot look at while sitting at a table, bring a clipboard with you so that you can still write down your observations.

Once you have picked your question, you will want to do research at the library and online. Chances are you know some things about your topic already, though. You might have learned a little about your topic in school. Maybe you learned through the observations you already made.

In a notebook, write a list of everything you know about your topic. Do not worry about exactly what you are writing. Just write down everything you know, from the big stuff to the small stuff.

SCIENCE TIPS

When you write your list, sit or stand near the thing or place you are observing. Being close to your topic might help remind you of things you already know.

After you have thought of everything you already know about your topic, you can ask your parents what they know about it.

HEADING TO THE LIBRARY

You can ask your parents to take you to a local public library or ask your teacher to take you to your school's library. Teachers and librarians can help you find the books you need.

A library's computer catalog tells you what books the library has on a topic, if they are checked in or checked out, and where to find them on the shelves.

Now it is time for research. The best place to start is the library. If you decide to find out why water forms on the outside of your glass, you will want to find books about water and weather.

When you get to the library, use the computer catalog to look up your books. Then search on the shelves. If you cannot figure out how to use the computer or if you have trouble finding the books, do not worry. The librarian can help you. You can even tell the librarian your topic and ask him to suggest some books for you to read. The librarian will tell you that the topic you are looking for is **condensation**.

Once you have found a book about your topic, you can take it home or start reading it right in the library!

GO ONLINE

After you visit the library, you can spend some time doing research online. If you have not used a computer for research before, ask your teacher or school librarian to help you.

To start, type some keywords into Google or another search engine. Then look at the links that come up. You want to read only good **sources**. Websites by schools, museums, and governments are good sources. These websites usually end in ".edu," ".org," or ".gov" instead of ".com." Websites of magazines like *National Geographic* are also good sources. Before you read the website, decide whether it will be a good source. Then, read carefully and write down useful information you find about your topic.

When researching online, you can write down your notes on paper. Doing this will help you put the research into your own words.

GET ORGANIZED

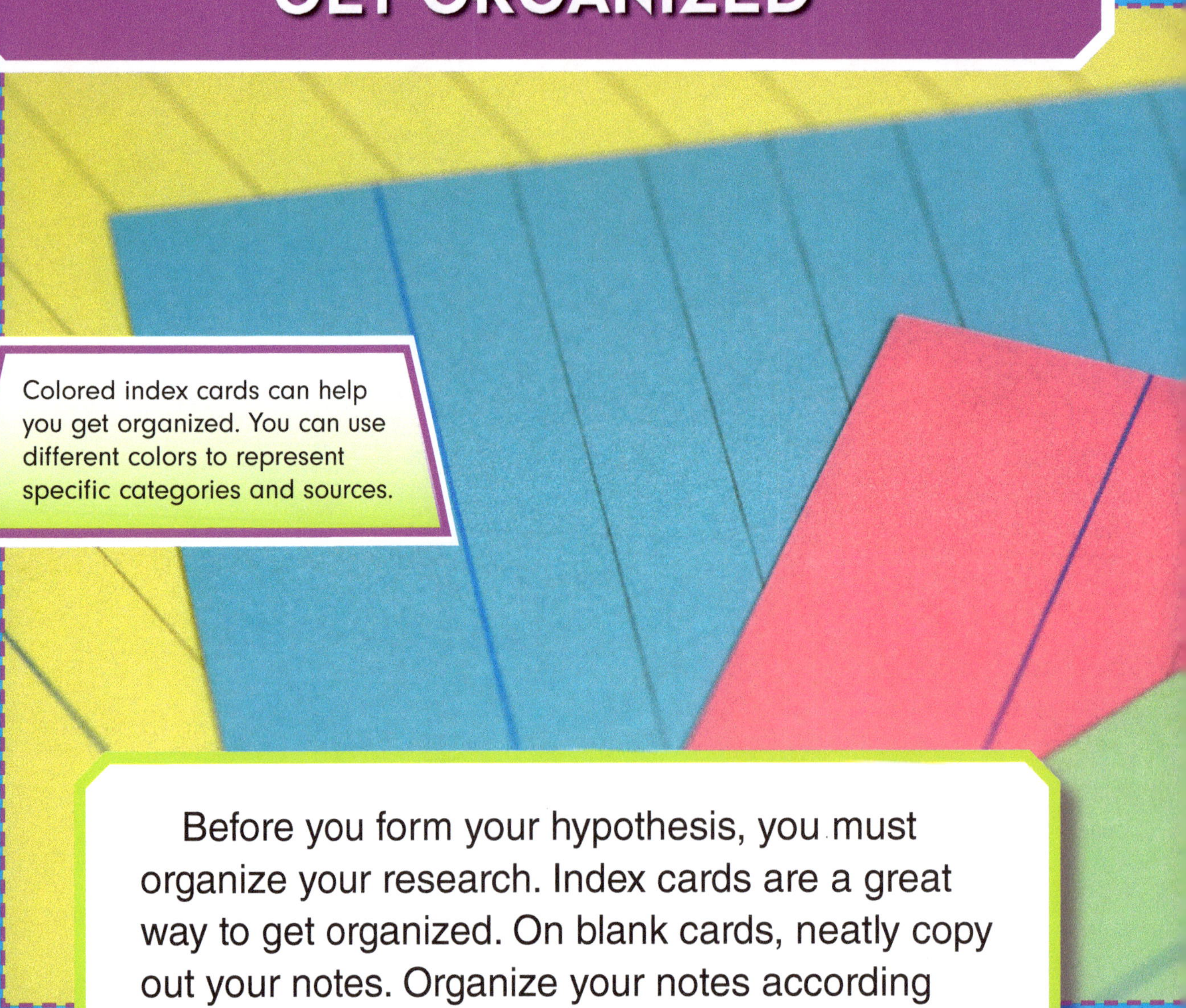

Colored index cards can help you get organized. You can use different colors to represent specific categories and sources.

Before you form your hypothesis, you must organize your research. Index cards are a great way to get organized. On blank cards, neatly copy out your notes. Organize your notes according to category. For example, put all the notes about condensation in general on one note card, all the notes about the water cycle on another, and so on.

Also, include any previous knowledge you have and the observations you made.

Next, get a highlighter and read everything very carefully. Highlight anything you wrote down that seems to be a clue to your hypothesis. Now you are ready to make your hypothesis!

Highlighting helps you find and see useful information. If you do not have a highlighter, you can use a lightly colored marker.

PREDICTING THE FUTURE

Now that you have done your research, you are ready to make a prediction. A prediction is what you think the answer to your question will be. You are going to use the knowledge you gained through your observations and research.

Perhaps your question is, "Why do water and fog form on the outside of a glass of ice water?" You might have learned that water forming on the outside of a glass is called condensation and is part of the water cycle. Your prediction then could be, "On a hot day, water will form on the outside of a glass of ice water but not on the outside of a glass of warm water."

Gathering all of your notes and reading them over before you start writing your prediction will help you write a strong prediction.

THAT GUESS IS EDUCATED!

Your hypothesis is an educated guess. An educated guess is a guess based on research and observations, as opposed to a wild guess, which is not.

Make your hypothesis as specific as you can. Take your prediction from the previous chapter

When thinking of your hypothesis, it can help to write down several different hypotheses and then choose the best one.

The glass of water experiment is a good experiment to do because you have easy access to glasses and water, which are the materials needed in the experiment.

a step further and say something like, "If a glass of ice water and a glass of warm water are put out on a hot day, the glass of ice water will become wet on the outside and the glass of warm water will not." There is your hypothesis!

SCIENCE TIPS

Do not forget that you will have to do an experiment soon to prove your hypothesis! Make sure your hypothesis predicts something you can really do an experiment about.

TRUE OR NOT? REVISING YOUR HYPOTHESIS

Of course, you do not know that your hypothesis will come true. It might not, and that happens all the time to scientists. You will know once you do your experiment because that is when you will test your hypothesis.

If your hypothesis does not come true, if both or neither of the glasses form water on the outside, then you may need to go back and **revise** your hypothesis. This does not mean you failed or that the experiment is over. In fact, it means just the opposite! Scientists sometimes have to revise their hypotheses dozens or even hundreds of times before they are able to prove them right.

If your hypothesis does not come true, try not to get frustrated. Revise your hypothesis, and experiment again. Remember that experimenting is fun!

DISCOVERIES CAN CHANGE THE WORLD

Scientists' work is never over, not even after their experiment is finished. They take the results of that experiment and use their new knowledge to form another hypothesis. Then, they are off experimenting again!

The scientific method is how scientists make discoveries that change how we understand our world. A doctor named Edward Jenner used the scientific method when he created the first **vaccine** to protect people against a terrible disease called smallpox. Jenner's discovery helped save millions of people's lives. Perhaps someday, you will make your own discoveries that help people.

To honor Edward Jenner for his work with vaccines, statues have been built of him, hospital wards have been named after him, and his house was turned into a small museum.

GLOSSARY

analyzing (A-nuh-lyz-ing) Examining something carefully and thinking about what it means.

condensation (kon-den-SAY-shun) Cooled gas that has turned into drops of liquid.

experiments (ik-SPER-uh-ments) Tests done on things to learn more about them.

hypothesis (hy-PAH-theh-ses) Something that is suggested to be true for the purpose of an experiment or argument.

laboratory (LA-bruh-tor-ee) A room in which scientists do tests.

observations (ahb-ser-VAY-shunz) Things that are seen or noticed.

researching (REE-serch-ing) Studying things carefully to find out more about them.

revise (rih-VYZ) To make changes to or improvements in something.

scientific method (sy-en-TIH-fik MEH-thud) The system of running experiments in science.

sources (SORS-ez) Things that give facts or knowledge.

specific (spih-SIH-fik) Stated in a way that is clearly and easily understood.

topics (TAH-piks) The subjects of pieces of writing.

vaccine (vak-SEEN) A shot that keeps a person from getting a certain sickness.

INDEX

WEBSITES

Due to the changing nature of Internet links, PowerKids Press has developed an online list of websites related to the subject of this book. This site is updated regularly. Please use this link to access the list:
www.powerkidslinks.com/smia/guess/